NO WALKS? NO WORRIES!

Maintaining wellbeing in dogs on restricted exercise

Siân Ryan & Helen Zulch
Photography by Peter Baumber

Hubble & Hattie

Hubble & Hattie

The Hubble & Hattie imprint was launched in 2009 and is named in memory of two very special Westie sisters owned by Veloce's proprietors.

Since the first book, many more have been added to the list, all with the same underlying objective: to be of real benefit to the species they cover, at the same time promoting compassion, understanding and respect between all animals (including human ones!)

Hubble & Hattie is the home of a range of books that cover all-things animal, produced to the same high quality of content and presentation as our motoring books, and offering the same great value for money.

www.hubbleandhattie.com

Disclaimer

Please note that no dog was deliberately frightened during photographic sessions. The images used to depict dogs feeling worried about specific situations were taken whilst the animals were exploring the novel environment of the studio setup. These images were later modified to enable us to illustrate the points we wished to make. Please also note that all body language is context-specific, and individual signals can mean different things in different contexts.

First published in September 2014 by Veloce Publishing Limited, Veloce House, Parkway Farm Business Park, Middle Farm Way, Poundbury, Dorchester, Dorset, DT1 3AR, England. Fax 01305 250479/e-mail info@hubbleandhattie.com/web www.hubbleandhattie.com. Reprinted February 2016. This edition published March 2019. ISBN: 978-1-787115-05-7 UPC: 6-36847-01505-3. © Siân Ryan, Helen Zulch, Peter Baumber and Veloce Publishing Ltd 2014, 2016 & 2019
Readers with ideas for books about animals, or animal-related topics, are invited to write to the editorial director of Veloce Publishing at the above address. British Library Cataloguing in Publication Data – A catalogue record for this book is available from the British Library. Typesetting, design and page make-up all by Veloce Publishing Ltd on Apple Mac. Printed and bound by CPI Group (UK) Ltd, Croydon, CR0 4YY.

Contents

Acknowledgements & Introduction

Acknowledgements

We wish to thank all the dogs and owners who participated in the photo shoots for this book: we could not have done it without you. We are grateful for your patience, willingness to work to get the perfect shot, and the fun we had whilst shooting. Special thanks also to Vicarage Veterinary Centre, Saxilby, and Whitehouse Canine Hydrotherapy Centre, Lincoln, for your expertise and use of your facilities.

Introduction

"You'll need to keep him quiet for the next few weeks ..." If you've heard these words from your vet, chances are, even if you're a very experienced dog owner, they caused immediate anxiety as you struggled to envisage how your eight-month-old bundle of Labrador energy, or your three-year-old mixed breed who accompanies you on a five-mile run every day, is going to take to this curbing of their activity. If that's the case, you are not alone: this is the reality for many dogs and their owners when a torn ligament, a degenerating joint, or some other illness or injury means that restricted activity is essential for recovery or pain management.

Experience in the field of problem behaviour has demonstrated to us that inappropriately managed restriction can have long term behavioural consequences for some dogs, and practical advice regarding appropriate management can be difficult to find. Therefore, the main aim of this book is to provide information for dog carers, as well as those working in the veterinary and paraveterinary fields, to help you and your dog manage, and even enjoy, a period of exercise restriction.

Other dogs may also benefit from the information in this book. For example, if your dog is older, you may have noticed that his physical capabilities have changed, and he may benefit from a shift in emphasis away from the physical to meet his needs for activity and enrichment. The book offers advice and support for his altered activity levels, too.

Likewise, you may have a dog for whom going on a walk is so stressful that his welfare is impacted by this daily outing, so

An off-lead romp provides your dog with mental and emotional, as well as physical, benefits.

4

Relaxing at home may be more important for your dog as he ages, although he still needs appropriate activities to meet his changing needs.

restricted walking could form part of a treatment plan for your dog to help him overcome his fears or frustrations. By using aspects of this book aimed at modifying your dog's activities you will boost his emotional and mental stimulation, which could help with long term behaviour modification, and improve his quality of life.

So what does walking mean to dogs?

Apart from the tiny minority of dogs who struggle to cope with walking, going on walks and engaging in active pastimes such as agility, tracking, or gundog work, means more to your dog than the simple physical exertion these involve. It's likely your dog has little control over his home environment: he may only be allowed out at times that you can manage; he may well only have access to certain areas in the house, and he may not have a view beyond the four walls if there are no dog-height windows. However, every time he goes out he is exposed to a range of social and environmental stimuli which can enrich his life, and, in addition, off-lead walks (and potentially carefully managed on-lead walks) give your dog a level of control over where he sniffs, what he investigates, and possibly even with whom he interacts – and control is important for wellbeing in all individuals.

Additionally, dogs are a social species and generally enjoy interaction with other friendly and appropriately behaved individuals – both dogs and people. This, too, contributes to a

Engaging in shared activities, such as relaxed sniffing, gives your dog the opportunity to be sociable, whilst at the same time deriving mental stimulation. If he is off-lead he has the additional benefit of controlling this activity and interaction himself.

positive emotional state, and thus positive welfare. For young animals, learning to build and maintain relationships with others is key to coping with meetings and interactions throughout life: therefore, restricted social interactions for a period of time can affect them in a number of ways on an ongoing basis.

From the foregoing it should be evident that restricted exercise may impact negatively on your dog in many aspects of his life. Whether the restriction is short or long term, removing the option to walk can have serious consequences, which need to be carefully addressed to ensure both short term wellbeing and long term quality of life.

What else happens when walking is restricted?

It is, of course, also important to remember that when activity is restricted for medical reasons, in most cases your dog will also be undergoing treatment for the underlying condition. Therefore, this book will also help you work with your dog to help him view necessary manipulations and treatments as positively as possible, reducing his distress as well as allowing him get the most benefit from the treatment; at the same time, making the professional's job easier and safer.

It is obvious how much pleasure both dog and person are getting from this encounter. Note how the dog is leaning in to the person's legs and stretching his neck for further contact. This is a dog for whom restricted access to social interactions could have a damaging impact on his mental and emotional wellbeing.

How to use this book

Keeping your dog appropriately occupied whilst on restricted physical activity can be challenging, requiring a re-think of your day-to-day routine, and adaptations in many areas of life for both you and your dog. The nature of the adjustments you might need to make will depend on the circumstances that have brought about the need for restricted activity, as well as your dog's personality and your current lifestyle. The starting point for getting the most out of this book – and best assisting your dog in his recovery or long term management – is to identify his specific needs.

Ask yourself the following questions before you go any further –

- How would you characterise your dog's general level of activity at this point? Is he a boisterous youngster who, at times, epitomises the notion of perpetual motion? Or is he a bit of a couch potato who prefers joining you in watching a DVD to exercising? Establishing this will help you get a handle on the level of alternative stimulation you will need to provide.
- How long will his restriction last? In other words, is this a short term problem where full activity can be resumed later? Or is this a long term problem with the need to make provision for a lifelong change in your dog's – and your – expectations?
- Does your dog have any behaviour problems, or is his personality such that he may be more likely to struggle with some of the aspects of a restricted lifestyle, medical management, or where some games and toys may be challenging? For example, if your dog is already sensitive to being handled, having dressings changed will be especially difficult, or if he doesn't like people near him when he eats, some of the games which utilise food in close proximity to people may not be suitable.

Additionally, dogs who have a tendency to view the world as a slightly scary place may, if restricted for a few months, struggle more with readjusting. Any specific needs like these should be worked into the rehabilitation programme.

Once you have given some thought to your situation, begin to work through the book chapters. The first goes into more detail regarding assessing your dog's needs, while chapter two will help you prepare for restricted exercise, should you have the luxury of time before a surgical procedure, or alternatively, give you some tips regarding managing in an emergency situation.

The next three chapters will help you to implement exercises and approaches to support your dog's physical, mental and emotional wellbeing, while the last chapter gives some ideas to help you return your dog to normal activity levels, should this be possible.

The information in this book will, we believe, give you skills that will help you to feel more in control, and thus able to cope with what we know is frequently a stressful and distressing situation. At the same time we hope that, through the information in this book, your dog will also feel more able to cope, and thus less distressed by his situation. Our intention is to promote the best possible welfare for dogs in what is potentially a difficult situation for them.

Please note that, for ease of reading, the text refers to the dog as a male throughout; however, female is implied at all times.

Something is this dog's environment is causing him concern. Note the tension across his brows, the focused expression, the nose lick, and the way he is tucking his tail and hindquarters under his body slightly. If you see similar body language from your dog, pay close attention to the environment that has prompted these signs. In this way, you can instigate immediate changes aimed at making him more comfortable, as well as build suitable management or training into his care plan to help him cope better in future.

Identifying your dog's specific needs

Your dog is an individual: no matter what breed, age, gender or colour he may be, or what you believe this will mean for his temperament and needs. He is unique. In addition, he has routines and expectations that result from his personality and his lifestyle, and which arise predominantly from all that he has experienced in the past. For this reason, when creating your dog's care plan it is important to base it on the dog you have; not on a generic model that proposes interventions based on breed or age, or a similar generalisation.

Having stressed his individuality, there are some character traits or categories of dog that may require specific considerations. For example, if your dog is under three years old – and especially if he is younger than 18 months – he may require different interventions to older dogs. Isolation and confinement as a puppy can have serious implications for behaviour as an adult, as he may miss out on essential socialisation or life experiences, or come to make negative associations with certain people or types of handling. When completing your care plan worksheet for your dog,

Sighthounds are not usually associated with agility-type activities; however, this dog loves all aspects of jumping, weaving and running over the contact equipment. His care plan should take into account the time he currently spends enjoying this training, and replace it with something suitable for him as an individual.

Remember!

Even if you feel your puppy has had lots of good experiences already, you should maintain this by ensuring your care plan contains suitable activities and excursions, as puppies – and even adolescent dogs – can quickly lose sociability or confidence through a period of isolation.

ensure you plan age-appropriate socialisation and experiences: generally at a maintenance level for older dogs, but with a higher level of exposure for youngsters. A young puppy can still meet people – and even other dogs (as long as these are calm and sociable) – throughout his period of restricted activity. In fact, it is important that he does so.

At the other end of the age spectrum, when creating a care plan for your older dog you may be considering his needs for

This adult dog is calm and relaxed when meeting the puppy on a lead, and is not engaging in a boisterous greeting. This makes him an ideal candidate for the puppy's continued socialisation, even when the pup has his activity restricted.

the first time since puppyhood. Whilst he may have needed and enjoyed long walks as a younger dog, this may no longer be the case, illness or injury aside. Age-related cognitive decline can be mitigated through provision of ongoing mental stimulation so, apart from replacing physical activity in the short term, increased mental activity can have long term benefits, too.

When assessing your dog to create his care plan, it is also important to consider how he acts when he doesn't get his expected levels of daily exercise. If, when he's well, he's whining and pacing for a walk, even before it's the usual time to go out, then he may take longer to adjust to new routines; requiring more time to learn how to relax and settle, as well as more mental exercise as his physical exercise declines. He may also benefit from increased variation in this new routine so he is less able to predict when things are going to happen, and therefore less likely to become excited in anticipation. In addition, consider and take into account your dog's specific traits or behaviours. Is he impulsive, or does he have low tolerance for not being able to do what he wants, when he wants to (is he easily frustrated)? Is he fearful in some situations, or does he dislike being handled, groomed or held? Does he already have an 'off switch,' or is he mentally and physically on the go all the time? All of these characteristics, and anything else you think of to describe his behaviour and routine, should be taken into account for his care plan.

For example –

- A dog who is impulsive and reacts without thinking may need extra work to teach him to sit or lie quietly in the face of exciting things, so that he doesn't move quickly and injure himself.

- A dog who is easily frustrated may need a slow approach to confinement training, and may benefit from the use of many distraction tools, such as food toys.

- If he doesn't like to be handled in certain ways and you know he will need to tolerate this, planning additional time for teaching this ability will be beneficial.

Also take into account any problems that sensory stimulation may cause him – for example, does he have any eye problems (which may influence your use of visual games and cues), or conversely, is he particularly stimulated by rapid movements or even the picture on the television, as this may rule out the use of certain types of activity.

Another important factor to consider is other dogs in the household, whose routines will almost certainly be affected by the change in your injured or ill dog, and his interactions with them will need to change, too. Consider how much time they spend together; the nature of their interactions (are there dogs with whom he curls up to sleep, and others he chooses to avoid or to play wildly with, for example?), and how you can maintain a safe level of contact and interaction between them. For example, do you need to accustom them to sleeping in separate rooms, or will time be well spent teaching them to participate in joint training sessions, where one dog is rewarded for lying quietly while you work on calm training activities with the other?

Hopefully, the above will provide a good starting point to plan your dog's care plan. Remember, too, that throughout his recovery period, or even if his restriction is to be permanent, it's important to reassess his needs as time passes, and adjust his care plan accordingly. The dog who found 'switching off and relaxing' impossible at first may be a Zen master within a few weeks, and you may therefore be able to reduce the amount of time you spend on relaxation exercises, and begin teaching other behaviours.

This dog is uncomfortable with the way she is being handled. Note the nose lick and the backward slant to her ears. Her weight is on her back feet and she is trying to pull away. There are several things which could be causing this discomfort: she may simply not like her feet being touched; it could be the proximity of the handler's face leaning towards her, or the arm over her back. She will need additional training to help her at least tolerate, if not enjoy, similar handling during her recuperation.

These two dogs enjoy playing together, and while the adult dog self-limits the speed and intensity of his play to match the puppy's ability, they would not be able to continue playing like this if either of them had to have their activity restricted. Consider ways in which you can provide alternative activities for similar friendships in your household or within your dog's social group while your dog is unable to play normally with his housemates or friends.

Remember!

Your dog may be more tired as you first shift the emphasis from physical exercise to mental stimulation and emotional support, as scentwork and free-shaping games, for example, seem to be particularly mentally tiring. Consider how much time you currently spend playing these kind of games, and his current level of ability and stamina for them. You can then plan how to allocate the time he spends playing games and training, and increase it appropriately over time.

No walks? No Worries!

Reading your dog's body language

Being able to read and respond appropriately to your dog's body language will help in both everyday life and during his period of restricted activity. After an injury or surgery he may be feeling sore or less sociable than usual, so it's important to respect his need for space as necessary. It's better to be cautious and avoid interaction rather than force your dog to accept your attention, or attention from another dog or person, and risk him resorting to aggressive behaviour in order to try to get you to understand that he wants to be left alone.

When you approach his crate or bed, if he makes eye contact with a happy relaxed face and soft eyes (you can see the muscles in the face are relaxed and the eyes not pulled wide or tight), and a softly wagging tail, go ahead and greet him. But if he stares with a tense face, turns his head away, appears stiff or shows other signs of potential discomfort such as lip licking or yawning (the latter may, of course, be because he has just woken up), leave him in peace in his bed or crate.

The same rules apply when he is interacting with another dog. If he shows any of the foregoing signs of discomfort, quickly but calmly move away the other animal before your dog is forced to escalate his signals to growling or snapping.

When making a fuss of your dog, give him plenty of opportunities to end the interaction. Every few seconds stop stroking him and wait for his reaction. If he asks for more fuss then carry on again; if he moves away slightly or stays still then respect his wishes and leave him be.

TIP

▶ *If you need to approach him to give him his medication or some other therapy, and he is signalling to you to stay away, call him and try to tempt him to you with a treat (if he can move from the place he is lying without assistance). If you are still struggling after a few attempts, it may be a good idea to chat to your vet, as your dog may benefit from additional medication to make him feel more comfortable.*

This dog does not want this interaction to continue. She has flattened back her ears, turned away slightly from her owner – and the nose lick is obvious. These are all indications that she would prefer to be left alone.

(Right and overleaf): The speed and intensity of the puppy's approach is worrying the adult dog. She has tucked her tail under her body and is keeping a close eye on what the puppy is going to do next; you can see the tension in her face. As the puppy approaches she stands up, but keeps all her weight shifted away from the puppy, remaining still and tense as the puppy sniffs her. Finally, she makes herself even smaller by lying down while the puppy remains interested in her.

If you see your dog experiencing a similar encounter to either of these two dogs, calmly and gently interrupt the interaction so that the worried dog does not continue to have an unpleasant encounter, and so that the greeting dog (in this case the puppy) does not continue to ignore the obvious body language from the other dog.

WORKSHEET 1: IDENTIFYING YOUR DOG'S SPECIFIC NEEDS

Name: .... Date: .... Age: ....

Primary carer: ....

Others involved in care: (eg family members, friends, dogwalker, etc): ....
....

Reason for restricted activity: ....

Time to prepare (yes/no)?: .... If yes, how long?: ....

Length of time activity to be restricted (eg 6 weeks, permanently, etc): ....

Current walks routine (number, length, type, on-/off-lead, etc): ....
....

Other activities: (training, car trips, games): ....

Relationships with other dogs/animals in the house: ....
....

Sociability (people and dogs) outside of the immediate household: ....
....

Special considerations (eg doesn't like being handled, scared of the vet, gets excited easily, easily frustrated, etc):
....
....
....

2 Preparing for a restricted lifestyle

If you know ahead of time that a period of restricted activity will be necessary for your dog, you have a chance to assess his needs, as described in the previous chapter, introduce altered routines, and/or implement training for new experiences and behaviours in advance. Both people and dogs can find change difficult to cope with, so making changes and establishing new behaviours ahead of time can help reduce stress for your family and your dog during the treatment period.

There will, of course, be times when it is impossible to prepare your dog and your family for changes in lifestyle. If your dog undergoes emergency surgery, or needs immediate crate rest because of illness or injury, for example, your focus has to be on his short term needs, with a back-up plan to make longer term changes as necessary. Specific advice for this situation is given throughout this chapter.

The chapter is divided into sections reflecting the different aspects of your dog's life, with examples under each heading –

- general points
- changes in lifestyle
- social interactions
- walking and activities
- handling and medical interventions

As you read through the chapter, consider how your dog's circumstances may have to be adjusted in each of these areas to create an individual care plan.

General points

To create your dog's care plan, at first simply list all of the things you need to change, introduce or remove from his daily life. Do not try to limit your options; just make a list of everything that seems relevant as you work your way through the book. When you have a list, prioritise the things that are most important to you and your dog, and rationalise the list so that it's achievable in the time you have available.

Feeding meals from food toys is one way to keep your dog occupied whilst he's confined to a crate. Food toys provide mental stimulation, an outlet for chewing behaviour, and give you a chance to take a break from his side if he is struggling with adjusting to restricted activity.

The shock of emergency surgery or confinement may leave you feeling helpless and out of control. One of the easiest ways to overcome this is to make a list of what you can do to help your dog when he first arrives home.

For example –

- Does your dog have a suitable bed, and/or an appropriate crate to come home to? Is his bed or crate situated in a cosy, draught-free spot where he can relax but is not unduly isolated from the family?

- Do you have a range of toys and chews that will help keep him occupied in the short term while he's learning to relax in his crate?

- Is there an easy, short and non-slip route from his crate or bed to the garden and other key places?

- Do you need to tape up the doorbell, or ask visitors to phone rather than use it in the short term to minimise the risk of over-excited behaviour immediately post-surgery?

Think about how your interactions with your dog are currently organised, and how that time allocation may need to change. For example, if your dog currently gets two, 40 minute walks a day, as you prepare him for restricted activity, first one and then both walks will be replaced with other activities and training. This is likely to also change the timing and length of his activity sessions so he still has 80 minutes of your time, but spread more evenly throughout the day.

When you create your care plan, tailored to him as an individual, choose a range of suitable activities from the suggestions throughout this book, and factor them into his daily routine so that on different days he has different challenges and interactions. In this way he is less likely to lose interest in any one activity but rather remain engaged and interested. When selecting activities ensure you take his mental, physical and emotional needs into consideration, and pick a range of activities which cater for all aspects of his wellbeing.

TIP
➤ *If your dog is walked by a dog walker, you will probably still need their help to interact with your dog at the established times, though rather than walking him you could discuss what other activities they can enjoy together instead.*

When prioritising, also consider what is going to have the biggest impact on your dog's quality of life and speed of recovery in terms of the changes that will be necessary. For example, if his activity is restricted because of surgery which requires him to be in a crate for long periods of time, then training him to feel safe, content and relaxed in the crate is essential, but this becomes less important or urgent for a dog whose activity is restricted because he is slowly aging or losing mobility.

Once you have your prioritised list, break it down into manageable chunks so that you can note when you achieve certain goals in your training, or implement changes that are required, so that progress can be seen. Your care plan can seem overwhelming, so ensure you have listed some simple changes first, such as installing a baby gate at the bottom of the stairs and getting used to closing it behind you, or setting up a crate and giving your dog his meals in there for a few days. This will help you to feel a sense of achievement, and thus more in control of your plan.

During the first few days of restricted activity your focus will probably be mainly emotional support for your dog. In the short term using chews, food toys or attention almost constantly to ensure he remains calm, relaxed and entertained is absolutely fine; you can gradually introduce longer periods of alone time or relaxation without distractions as he settles and adjusts to being confined. In the same way, using food treats to distract him during procedures or handling that he may not like is absolutely fine in the short term, or in an emergency situation. Ideally, eventually wean him off distractions and teach him to tolerate (or even enjoy) these procedures, and only use food, attention or a toy as a reward.

During these early days you may need help to keep him occupied, and assist in managing, treating and moving him, but additional people coming and going may be unsettling for him (unless he is used to this in normal daily life), and may also make teaching him to relax on his own harder in the long term, so do try to balance this with alone time. Also, if he has residual pain and discomfort, he may not want physical proximity or attention from anyone – or at least those he doesn't know so well – so remember to pay attention to and respect the signals he is giving, and adjust your behaviour or advise others accordingly.

Predictability and control are important to your dog, especially when he has lost his usual ability to make decisions and exercise some control over his environment. If he is normally able to go anywhere, sleep anywhere, or interact with people or other dogs with a degree of autonomy, adjusting to losing that control can lead to emotional distress, which can be seen in a range of attention-seeking or potentially aggressive behaviours. Give him as much choice and control as you can: for example, allow him to choose if he wants you to make a fuss of him, and don't wake him if he's snoozing.

continued page 20

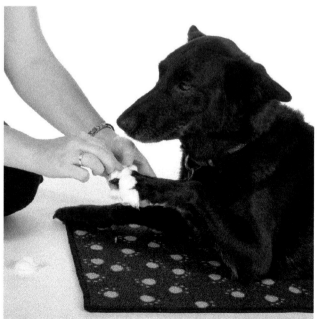

Understanding what is going to happen next increases predictability and thus decreases stress. For example, if he knows that when you put a mat on the floor and ask him to lie on it, it means that you are going to change his dressing and that a treat will follow, he will be more likely to relax and co-operate. For this reason, as soon as possible, think about the things he needs to learn, and begin to teach him these so that any handling or care he requires will be positive experiences.

Begin by feeding him for lying on the mat at first, and then progress to introducing the dressing or bandage before his operation if you have the opportunity to do so.

Sometimes the mat can become a predictor of a negative experience for your dog, and he will try to avoid or move away from it as you place it on the floor. The best way to prevent this is to spend a lot of time feeding him treats on the mat before you start any handling training, so that he has a long history of good experiences to buffer any negative ones in the future. If this is not possible, continue to use the mat for the procedures you need to do, but save special, very high value treats to use as a distracter. The presence of the mat continues to give him predictability, and he knows that at other times when the mat is not present nothing unpleasant is going to happen to him.

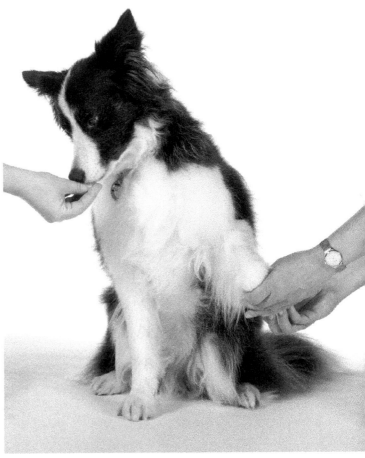

If you do not have sufficient time to teach your dog to co-operate with handling procedures, use treats to distract him while you check his wound or perform other necessary handling.

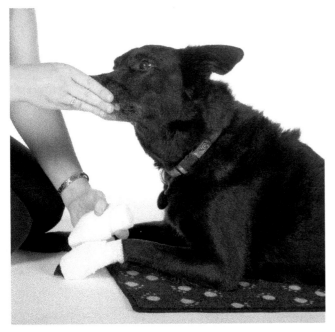

No walks? No worries!

Lifestyle changes

One of the first things to think about when preparing for your dog's reduced activity is the impact it will have on your home. If he currently has unrestricted access to the whole house, sofas and beds, having all that taken away at once may be difficult for him – and you – to cope with. Some dogs will accept this change easily, whilst others may become frustrated (which may be seen as aggressive behaviour), depressed or anxious. For this reason, if you have time to prepare him, gradually, in a staged manner, restrict his access to areas of the house he won't be allowed into.

It's particularly important to consider where he currently sleeps, and what changes will be necessary to ensure he's comfortable with any new arrangements. Will you need baby gates to prevent him going upstairs, or do you have a suitable crate if

he requires total confinement for a period? See the appendix on relaxation and crate training for advice on how to introduce a crate to ensure he's happy and comfortable inside. Another aspect to consider is visitors. Some dogs will react energetically to the doorbell/doorknocker, and you can either spend time habituating or desensitising your dog to the noise – by repeatedly ringing the

Sleeping on chairs is likely to be forbidden for the duration of your dog's restricted activity because of the risk of injury from jumping up or down.

If you normally cuddle with your dog on the sofa, begin to sit on the floor with him instead, so that he still gets the closeness which you both enjoy but without him jumping up.

If he is used to jumping up when he greets people, it will be beneficial to teach him to keep all four paws on the floor (use treats to focus his attention on the floor when people arrive, or teach him to sit for greeting), and ask visitors to bend down to greet him instead.

The best long term approach to having your dog greet people calmly at the door (which is useful for dogs at all stages of their lives), but which takes time to implement, is to teach him to lie on a mat, in his bed or in his crate when the doorbell rings. As well as being a useful exercise, the training involved provides mental stimulation, rewards calm behaviour, and helps to reduce frustration when he may later have to be in his crate in the presence of exciting stimuli.

It may be necessary to keep him on a harness and lead in the early stages of training so that you can manage him, and prevent him practising jumping up.

bell until he no longer responds, and then rewarding him for being calm – or, if you do not have time for this, deactivate it and attach a note to your door asking visitors to call you on arrival instead.

Social interactions

As well as getting down on the floor to cuddle your dog (as shown in the image above), you may need to introduce some other rules for interactions with family and friends. It's important that your dog continues to have social interactions (see Chapter 5), but everyone needs to be aware of how and when to greet or fuss him.

Introduce a harness or head collar as soon as possible so that your dog is used to wearing one before he takes his first walk during recovery. Introduce it as described in the sequence of photos for muzzle training (page 27) so that your dog associates it with something positive, and is comfortable wearing it. If you do not have the time to introduce it slowly, use a treat to tempt your dog to place his head inside the headcollar or harness, and immediately do it up and head out for a walk. If he tries to fight it or remove it, gently distract him and continue walking, using treats, if necessary, to continue moving forward.

you manage him so he won't cause himself further injury when allowed outside.

Taking your dog for rides in the car – as long as he enjoys car travel – can be an excellent form of mental stimulation for him.

You may need to discuss and agree new house rules with family members or frequent visitors, ensuring that everyone is able to read and respond appropriately to your dog's body language when interacting with him.

Walking and activities

Think about what your walks are like at the moment. Does your dog leap around excitedly at the sight of the lead? Once outside, are your walks a mixture of calm, loose lead walking with plenty of opportunities for sniffing and interactions, as well as some off-lead time; are they fully off-lead, apart from five minutes of stressful pulling to get to and from the park, or are they a constant battle of pulling? You may need to spend time habituating him to the sight of the lead by leaving it lying around all the time, regularly picking it up and putting it down again until he no longer reacts. If he pulls hard when on-lead, consider whether you need to teach him how to walk calmly on a lead (see Appendix 1 *Loose lead walking* for further information), or if you have not had time to prepare, you may need to source a suitable harness or headcollar that will help

If a ramp will be necessary because your dog is unable to jump into the car for a period of time, before attaching it to the back of the car, introduce your dog to it while it is on the floor. Place treats along it so that he learns to walk on it and is comfortable doing so.

Ensure you support your dog when he is using the ramp, and stay close to him in case he should slip or try to jump off. Use food as necessary to reward him.

If your care plan includes car journeys – even if just to the vet for post-surgery follow-up – give some thought to how your dog will get in and out of the car. Is he happy to be lifted (can you lift him safely and comfortably?), or do you need to practise this? Would a dog ramp of some sort be what's needed, and how do you teach him to use it, if so?

If your dog is used to doing regular activities or training classes, prioritise replacing them with suitable mental and physical activities at home, or consider ways he can still attend while remaining calm and without risk of further injury during his recuperation.

Handling for veterinary procedures

It's likely your dog will be handled by your vet and your family far more during his treatment than he would be normally. If you are aware that your dog does not like to be touched on cerrtain areas

No walks? No Worries!

To help your dog learn to tolerate or enjoy being handled, begin by gently touching an area of the body that he usually doesn't mind being touched, and follow the touch with a food treat. Repeat a few times so that your dog understands that touching equals a treat. Repeat this on different areas of the body. If he has a sensitive spot that you will really need to touch, start by touching a place close to the spot, and, over time, move very slightly closer to the sensitive spot, rewarding after the touch every time. Continue to do this, in short sessions, gradually moving closer to the sensitive spot until he is happy to be touched there. If necessary, you can use the same process to build the ability to not just touch, but to pick up, squeeze or bandage as necessary.

Hydrotherapy will inevitably involve your dog getting wet, so teaching him to enjoy being towel dried or dried with a hair dryer is helpful.

room and feed treats to your dog to help him feel better about being there: this will help buffer him against those times when he goes in for less pleasant experiences.

Many dogs will benefit from hydrotherapy and/or physiotherapy as part of their physical and mental stimulation and recovery. Discuss the options with your vet, and, if possible, take your dog to meet his therapist in advance of treatment. This will allow you to introduce him to the noises, smells and equipment involved, and help you decide which behaviours it may be beneficial to teach him in advance: for example, standing for physiotherapy manipulation, or lying down to enable appropriate stretching.

Your dog may need to become accustomed to new equipment, such as an Elizabethan or inflatable collar to prevent him licking stitches or dressings; boots to prevent dressings becoming wet when outside; a harness or head collar for walks, or a muzzle for safety during examination and treatment.

of his body, allow time in your care plan to work on changing his attitude. It will make veterinary visits and treatments a lot less stressful for everyone, including your dog. It can also be useful to make some trips to your veterinary clinic to simply sit in the waiting

A range of different collars are available which are designed to prevent your dog from being able to lick or chew at his stitches or dressings. When choosing the most appropriate one for your needs, bear in mind the length of your dog's neck or nose, and the location of any wound. The traditional Elizabethan collar will enclose your dog's head like a lampshade, but many dogs find it uncomfortable to wear for prolonged periods, and it may prevent him from being able to eat or drink. The inflatable collar pictured here may give less protection to some wound sites but is frequently tolerated better. It is important to discuss these options with your vet.

Such equipment can be introduced using the same general principles as described in *Muzzle training* (below): short exposures to begin with, increasing the length of time he wears the equipment at your dog's pace, and pairing it with food rewards to ensure a positive experience.

TIP
➤ *When using food for training rewards, take this from his daily food allowance and reduce what goes in his bowl or food toys accordingly. Bear in mind that during his recuperation he will use a lot less energy than normal, so his diet should be adjusted to reflect this.*

REMEMBER!
Use your dog's favourite reward to teach him and manage his behaviour. Food is usually the easiest and most practical, especially when active games such as tug or retrieve are not possible, but if your dog really does value fuss or grooming, for example, use these to reward him.

Muzzle training
You can use the technique described in the following images and captions to teach your dog to wear any piece of equipment, such as a harness, headcollar or protective boot.

continued page 34

A basket muzzle is the preferred choice if your dog needs to wear a muzzle for anything other than a short period of time or specific treatment. Basket muzzles, as pictured, allow him to pant, drink and take treats, unlike cloth muzzles which prevent his mouth from opening properly. Present the muzzle (or other piece of equipment) and feed a treat immediately after you have presented it. Repeat this a few times and you will see him begin to anticipate the arrival of food when he sees the muzzle.

At this stage, place a few treats inside the muzzle to encourage him to place his nose inside to get the treats.

Gently remove the muzzle from his nose while he is still eating the treats inside it, so he does not learn that he can back out of it himself.

After a while offer the empty muzzle, and, as soon as he puts his nose in, feed him a number of treats through the end, once again removing it gently before he can back out.

Progress slowly to being able to lift the straps around the back of his head, and eventually to fastening them. When he has the muzzle on with the straps fastened, feed him just for wearing it.

Begin to leave it on for longer periods, until he is wearing it for several hours a day in advance of his surgery. If he starts to scratch at or try to remove the muzzle, distract him and only remove it when he has stopped trying to get it off.

Right: If you need to muzzle your dog in an emergency situation without prior training, if at all possible use a different muzzle for the emergency situation to that you use for your long term, slow acceptance training. For example, use a cloth muzzle for the emergency situation and a basket muzzle for the long term training. Use a high value treat (such as a tube of squeezy cheese or pâté) to encourage him to place his nose inside the muzzle, and continue to feed him treats while you fasten it. No matter what your dog's reaction, try to keep calm at all times, and, whatever you do, do not reprimand him if he fights the situation.

Important note: DO NOT leave your dog wearing a cloth muzzle for long periods of time, and never leave it on when he is unsupervised as he cannot drink, pant or eat properly whilst wearing it.

No walks? No worries!

Now that you have given some thought to the changes and/or equipment you may need to introduce, and any new behaviours you have to teach your dog, the following chapters give suggestions for the kinds of activities that will help support your dog physically, mentally, and emotionally during his period of restricted activity.

Remember!

Many of the changes necessary to prepare your dog for the inevitable consequences of his physical restriction involve using food to manage, entice or reward appropriate behaviours. Excess weight can be detrimental to recovery from surgery, as well as the long term health of joints, so treats should be carefully balanced against daily calorie intake. For many dogs, it can also be beneficial for them to lose weight in advance of surgery. Discuss your dog's weight with your vet if you are not sure whether or not it is ideal.

Discuss your dog's weight loss needs with your vet and consider changing his diet and the amount of food he receives in advance of surgery. In addition, if you know your dog will struggle to cope with reduced portion size on a weight loss programme, chat to your vet about safe ways to bulk out his food: for example, through the addition of chopped raw carrots or broccoli stalks.

WORKSHEET 2: PREPARING FOR A RESTRICTED LIFESTYLE

What changes need to be made within our home?
(eg: we need to install a stair gate at the foot of the stairs) ....

....

What new equipment does he need to learn to tolerate?
(eg: I need to teach him to wear a protective boot) ....

....

How will I amend his daily routines (social interactions, walks, other activities)?
Current routine is *(eg: two hour-long walks every day)* ....

....

End goal and date by which I would like to achieve this
(eg: 3 x 10 minute training sessions a day, plus 2 x 20 minutes of scent games, etc) ....

....

What handling is he going to need?
(eg: I will have to get him into and out of the car to go to the
vet and physiotherapist, so need to teach him to use a car ramp) ....

....

What other considerations are there?
(eg: he is currently at his optimum weight so I will have
to monitor his calorie intake as I begin to use treats more often) ....

....

Monitoring progress and amending plans
(eg: he has learnt to lie quietly in his crate with me in
the room; I must now start leaving him alone for short periods) ....

....

....

3 Managing your dog's physical needs

From the perspective of physical activity, restriction will mean different things to different dogs, and how this should be implemented will depend on many factors. A couple of aspects which will influence your approach are –

- the duration of the problem
- the severity of the problem

For example, restriction may involve several weeks of total crate rest before returning to normal activity, during which time he may only be allowed out of the crate under strict supervision to go to the toilet, returning directly to his crate. He may even require physical support whilst going to the toilet because of the nature of his illness or injury.

For another dog, restricted activity may mean a long term reduction in the length or frequency of walks, or a permanent change in lifestyle such as a ban on jumping on the sofa or going up stairs without help. For this reason, when reading this chapter, consider your dog's specific situation and what this may mean for your planning.

In this chapter we cover –

- the use of the crate
- moving dogs who need assistance
- things to bear in mind when going for walks
- using the space you have for physical activity
- a few words about hydro- and physiotherapy

The crate

Whatever the nature of the restriction your dog faces, creating a limited, safe, secure space for him, where he feels comfortable and relaxed, is a key part of ensuring you are able to successfully complete the treatment programme your vet has prescribed. Using a crate, a puppy pen, or a room subdivided with a barrier to restrict him to a small space so as to limit his movement, is the easiest way to ensure that your dog cannot do anything he is not supposed to

during his recuperation. It also means you do not have to supervise him 24 hours a day, but can relax in the knowledge that he is safe (provided he is not going to further injure or distress himself trying to break out). Correctly introducing your dog to spending time in a restricted space is very important to ensure that he is relaxed and comfortable there. Please see the crate training appendix for information regarding introducing your dog to the crate to prevent frustration-related behaviour.

Where possible, and if appropriate after discussion with your vet, give your dog supervised time out of his crate during the day. If possible, observe his natural sleep and wakefulness patterns, and try to time any periods out of the crate with when he is naturally more restless.

All but the shortest of walks out to the toilet may be off limits, but lying in the garden or going for a drive in the car may be possible. Do not underestimate the benefits for your dog of a simple change of scenery, which comes, of course, with accompanying smells.

A canine physiotherapist may be able to give you ideas regarding gentle massage or manipulations which your dog may enjoy. However, this must be undertaken in consultation with your vet so that nothing is done that may be detrimental to his recovery.

Even if he cannot move around much, if your dog enjoys physical closeness, allow him out of the crate to sit on the floor with you and your family or friends.

Depending on the nature of your dog's illness or wound, it may be easier to physically support him when he is moving around, as well as prevent any unwanted activity, by using a harness and lead to manage him when he is outside the crate. There is a very small risk with any collar or harness that they can become caught in the bars of the crate if worn inside the crate, but this should be balanced against the possible difficulty of fitting the harness as he leaves the crate: decide what is best for your dog and your circumstances. If he becomes very excited when he sees his lead it may be easier to keep a very short lead (really, just a 'handle')

Left: You can use time out of the crate to groom him or practice some handling exercises – teaching him to enjoy having his nails clipped or his ears cleaned, for example.

attached to his harness at all times so that you can manage him out of his crate without needing to bring out his usual lead.

Ways of keeping your dog appropriately occupied whilst in the crate are covered in the chapter on mental stimulation.

Assisting your dog to move

As mentioned previously, some dogs may not be able to walk at all in the acute recovery stage, so lifting and carrying them may be necessary. Ask your vet to demonstrate how to safely (for both you and him) move your dog when you need to, as specific injuries and surgeries sometimes benefit from specific techniques. The images on pages 38 and 39 show some lifting techniques which are generally safe.

continued page 40

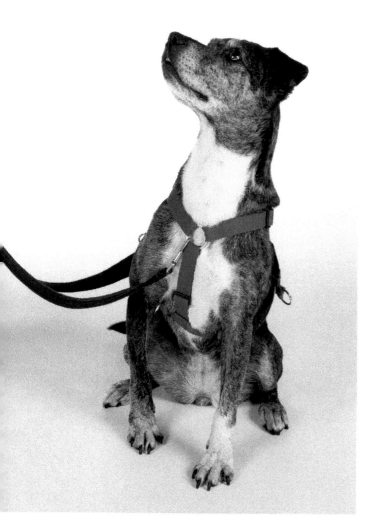

If you decide that your dog will wear his harness at all times, ensure it fits appropriately, does not rub or pinch, and is well padded to protect delicate skin. It should sit well away from any wound sites or stitches. Harnesses which provide multiple points for lead attachment can give you greater control over his movement by using a double-ended lead, or two separate leads, attached both front and back. Clipping the lead to a waist belt so you have both hands free for lifting or support can also be beneficial.

If your dog cannot move himself at certain times, then lifting and supporting him appropriately is really important. In addition, being able to physically move a dog who cannot walk can provide more opportunities to give him appropriate physical and mental stimulation, and emotional support.

When lifting a dog ensure you support him fully, keeping his body close to yours, and your hands or arms well away from any wounds. Discuss with your vet if a specific harness may be more suitable for supporting your dog, given his particular requirements, or if he does not like to be lifted and carried.

Another way to help your dog when he is not very physically able is by minimising slippery surfaces around your home. If you have lots of tiled or laminate floors, getting around may be difficult for him, and strategically placing a few inexpensive mats or runners next to his crate or bed (so that he can get out easily), as well as on the main routes that he uses to walk around the home, can greatly benefit him.

Using your space to help your dog

With a little imagination you can turn your house into an exercise area for a recuperating or long term activity-restricted dog. Take into account all of your rooms as well as your outside space, and how you can use this effectively.

Some examples are –
- Practise slow, loose lead walking both indoors and outdoors, using treats to reward him for being alongside you. The slow pace will encourage calmness, and minimise risk of injury; it's also useful practice for when he's allowed to go for walks again.

See the loose lead walking appendix for advice on how to teach your dog how to walk alongside you.

- Another useful trick to teach him whilst he's restricted is to move behind you when asked, and/or take refuge behind a pop-up umbrella, as both these manoeuvres can help to minimise unwanted attention from and contact with other dogs when he is out and about at a later stage.

- Engage him in some of the scent games described in Chapter 4 *All about mental stimulation.*

Teaching your dog to move behind you before you start going for walks again is useful, as it will enable you to give him greater protection if dogs or people start paying him attention when the interaction is not appropriate for him. Use a piece of food to tempt him behind you a few times, rewarding him when he is in place. After 4 or 5 repetitions, use the same hand movement but without food this time: he should still follow your hand and you can reward him as soon as he moves behind you. When he is happily and promptly following your hand without food, you may want to add the verbal cue 'behind' immediately before making the hand movement, and then gradually fade the hand movement so he moves behind you on verbal cue only. In an emergency situation you can always entice him with food as necessary.

For even greater protection on walks, combine the move behind with opening a pop-up umbrella, which will screen your dog from the unwanted approach of another dog, and can help persuade them to leave you alone when their owner is out of sight or unable to call them back. Allow your dog plenty of time to investigate the open umbrella before combining it with the move behind, and then teach your dog that opening the umbrella means treats for him so that he is not worried when it appears. Open the umbrella, feed your dog a treat, and then ask him to move behind. Repeat a few times and, as long as you can see he is unconcerned by the umbrella opening, feed the treat only after he has moved behind you when asked.

- If you have a passageway, a room or area of the garden that is suitable, construct a safe obstacle course by arranging objects such as large cardboard boxes or a broom lying on the ground, to encourage him to step around or walk carefully over items, thus exercising his muscles and joints in different ways.

If you decide to create small obstacle courses, check that these are appropriate for him by discussing with your vet. If your dog becomes overexcited when using them, manage him on a harness and lead so that he walks slowly and deliberately. Use treats or scattered food as necessary to keep his attention on the ground, checking he does not become so focused on the food that he fails to pay attention to what his legs and paws are doing.

All of these exercises can be practised on or off the lead, depending on what is most appropriate for your dog at the time.

Going for walks

When able to resume walking your dog outside again, pick your route with care. Think about the surface underfoot –pavements may be better than grass to avoid slipping, but grass or soft sand is better to prevent concussion on joints, for example – and how busy your route might be. Whilst an area with lots of doggy traffic will provide interesting smells, the risk of another dog bothering yours may be high. Equally, the sight of other dogs may cause your dog to become overexcited or frustrated. If either of these situations occur, use your pop-up umbrella (if he is comfortable with it) to provide a screen between you and the other dogs, and move him away as quietly and calmly as possible.

Tip
❥ *Don't be afraid of politely asking other dog walkers to call their*

dogs to them before they reach yours, or telling others that your dog is unable to greet them at the moment. Be sensible, though: don't take him somewhere all the other dogs will be running around off-lead and expect other owners to keep their dogs away from him.

Another thing to bear in mind is how far he might be able to walk initially, and whether you could safely carry him back to the car or home if he gets tired. It is far better to walk back and forth a few times across a short distance than to attempt a walk round the block, and end up having to carry him home. The initial excitement at being outside after days or weeks of confinement may encourage him to do more than he can actually cope with, and lead to soreness or stiffness as a result. Less is more as he builds his muscles and stamina again. Draw up a plan of gradually-lengthening walks and stick to it, even if he appears capable of more.

Left: A good hydrotherapist will encourage you to familiarise your dog with their equipment, such as the treadmill pictured here, prior to filling it with water. This enables your dog to gain confidence and therefore derive more benefit from his session, as he will be more relaxed.

Above: The underwater treadmill is filled to the correct height for your dog and his specific problem. Your hydrotherapist will support him as necessary during the treatment.

Hydrotherapy and physiotherapy

Discuss with your vet the possibility of referring your dog for hydrotherapy and/or physiotherapy. Both can provide invaluable additional support for healing and recovery, as well as an outlet for his physical energy. A suitably qualified hydrotherapist will work alongside your vet and physiotherapist to assess your dog, and help with his recovery. If you have been able to introduce your dog to the therapists, environment and equipment prior to any treatment, he will be able to get the most from his sessions once given the go-ahead to begin.

A physiotherapist, as well as helping your dog to get the most from his water sessions, should he be going to hydrotherapy, will advise about additional bends, stretches and exercises he can do at home. These will also form part of his safe physical activities when he's outside the crate.

Left: For some dogs, the pool will be more appropriate for exercise and therapy. The hydrotherapist will be in the water with your dog at all times, and will encourage him to swim in the most appropriate manner for his injury/condition.

This is only an illustration of the sort of stretch exercise your physiotherapist may recommend. Specific exercises, tailored to your dog, can help him recover more quickly, and protect him from further injury in the future.

No walks? No Worries!

Things we need to put in place. *(eg: buy a crate; move mats to create non-slip routes through the house)*

....

....

Things we can teach him *(eg: to be comfortable wearing a head collar; to move behind me when I ask him to)*

....

....

Activity planning *(eg: week 1: spending time sitting in the garden for half an hour every day)*

Week 1: ....

Week 2: ....

Week 3: ....

Week 4: ....

Week 5: ....

Week 6: ....

Week 7: ....

Week 8: ....

Notes regarding ongoing progress, problems and changes
(eg: he is trying to move too quickly when he comes out of the crate so I need to slow him down using treats)

....

....

....

As well as providing physical exercise, taking your dog for a walk gives him access to a world of sniffs and smells, social interactions, learning opportunities, and visual stimulation, all of which are important for his mental wellbeing. When physical exercise is limited, so, too, are all those other important aspects, which can negatively impact on your dog's welfare in both the short term and, potentially, with lasting consequences. With a little forward planning it's easy to replace the time spent walking your dog with a range of activities that will provide him with the stimulation he needs, and thus help maintain his mental wellbeing during recovery.

Hints and tips for types of activities
FOOD TOYS
Whatever toy you choose, to get the most benefit from using it your dog will need to learn how it works. Always introduce the toy on the easiest setting, and with some high value treats inside. You can also try smearing it with a little liver or cheese paste at first to make it really attractive to him. Encourage your dog to investigate and work at the toy, and ensure he is getting at the food quickly

continued page 51

Although many dogs enjoy playing training games for a toy reward, such as a tug game, when exercise is restricted it is often difficult to use play in this manner. For this reason most of the suggestions in this chapter revolve around food use, and we suggest that, instead of feeding your dog from a bowl twice a day, you use his daily portions of food to fill food toys, play interactive games with him, as training rewards, and as search items for him to find. You may need to be creative in what you use as food treats: for example, if your dog is on a weight-loss programme low calorie treats may be needed (eg natural popcorn or pieces of apple or carrot), or if you have a very fussy dog, you may have to try a range of high value treats such as ham or cheese, reducing the amount of his usual food to compensate for the extra calories. For dogs with special dietary requirements, discuss safe treats with your vet.

Many different food toys exist; some – such as the original Kong™ – encourage dogs to chew or lick to remove the food; others, like the Twist 'n Treat™ shown here, roll or tip over, and require some manipulation by the dog in order to release the food. Choose ones that will be suitable for your dog and his current physical abilities.

Using a range of different toys will encourage your dog to try different tactics and strategies to remove the food. This can be particularly useful for dogs who become easily frustrated as it encourages quiet perseverance in working for a delayed reward.

Empty food containers such as cereal boxes, egg cartons, tea-bag boxes, and toothpaste tube boxes (with all metal clips and sticky tape removed) make great food toys. Stuff them with some waste paper and kibble or treats, close the box and give it to your dog to rip apart and get to his dinner. You can also use old envelopes, but be aware that you might end up with a dog who regards all parcels and envelopes as toys for him to destroy, so don't leave your post lying around!

When introducing your dog to interactive toys, as with food toys he plays with on his own, make it easy for him to gain the rewards at first. Most games come with full instructions on how to do this, but general principles include ensuring that the game is on its easiest setting; only partially hiding the food in the various compartments, and encouraging your dog to move the pieces with his nose or paw as appropriate.

enough to prevent him becoming frustrated and abandoning the toy.

As he becomes more proficient, make it harder for him: by using a mixture of wet and dry food, for example, or by using more difficult settings, depending on the toy. Original Kongs™ can be filled with wet food (for example, kibble soaked in dog gravy), and frozen for a particularly time-consuming and challenging treat.

Plastic bottles containing treats or kibble can be entertaining for him to roll around and empty, but never leave him alone with these, and always remove it if he starts to rip the plastic, as he could cut himself or swallow the pieces.

INTERACTIVE GAMES

As well as food toys that your dog can play with on his own, there are several interactive games that you and he can play together. These games, such as those in the Nina Ottosson™ range, vary in complexity, and will encourage your dog to problem-solve in order to reach the food reward. Using his brain in this way can be very mentally tiring, so keep play sessions short at first while your dog is learning how to solve the puzzle.

SCENT GAMES

Dogs have a very well developed sense of smell, and love to use it

Some dogs will try to use sheer strength or rapid scratching at the toy in order to open the compartments, often with some success. However, in order for these toys to be most beneficial in providing mental stimulation, you want your dog to approach them with control. To encourage this, wait for him to be calm before putting the toy on the floor, and remove it if he starts to scratch too vigorously or bite at it. Give the toy back again immediately he calms down, so he begins to realise that being calm is what gains him access to the toy. Working together with the toy in this manner can be a good substitute for the more active games that you can no longer enjoy together.

REMEMBER!

There is no such thing as an indestructible toy, and even tough toys should be supervised the first few times you give them to your dog to ensure he can't break off bits and swallow them. Also, replace old, damaged toys that may break into pieces more easily.

to find food, toys, and even people. Much of their time on walks is spent sniffing at the scents left behind by other dogs, people and animals, and building a scent picture of events since they were last in that area.

You can play many different types of scent games with your dog: take a look at the pictures in this chapter for some inspiration.

In addition, games such as 'hide-and-seek' can be played with your dog by hiding treats or filled food toys around the house for him to find. At first, have someone restrain him gently, or ask him to sit and wait (if he knows how to), and let him watch you hide the item (behind a piece of furniture or under a cushion are good hiding places to start with). As you release him to find the toy, say 'find it' and let him go and get it.

After a few repetitions, begin to shut him out of the room before you hide the toy, and use 'find it' as you open the door to encourage him to search for it. When he gets really proficient at this, hide the toy in different rooms, or on different floors (assuming it's appropriate for him to climb stairs), for him to look for and find.

REMEMBER!

If your dog is uncomfortable eating in close proximity to you, or defends his food if you approach or try to remove it, then do not use the interactive food toys with him. Carefully manage all interactions that involve food so as to avoid problems.

To give your dog an opportunity for some sniffing, the easiest thing to do at home is to scatter his dry food across your lawn, kitchen floor or mat so he can spend time finding it. At first, make it relatively easy by creating small piles, especially if your grass is long, but soon you can scatter it randomly and leave him to spend several minutes looking for and eating his dinner. You can do the same with moist food by placing small portions in a selection of small bowls that you spread around or hide.

You can also use cardboard boxes for scent games. Hide a piece of food in one box and then place the box in amongst several others scattered on the floor. Encourage your dog to 'find it,' and allow him to investigate the boxes until he finds the food.

Flower pots, muffin tins, dustpans and other similar objects are great for hiding food in and under, encouraging interactive games. You can teach your dog to knock over the one food- or toy-containing flower pot from a selection of two or three in front of him, or you can hide several pieces of food in different objects and encourage him to 'find it' by investigating them all.

You can also begin to add multiple treats or different items (such as a favourite toy) to the search, and, if he likes bringing things to you, turn the activity into a search, find and retrieve game. Reward him for bringing things back with a game played with the toy he has found, if it's appropriate to do so.

Another game most dogs enjoy is searching for a member of the family or a friend, instead of a treat or toy. Reward him with a fuss or a tasty treat for finding the hidden person.

TIP
♦ *If you find that these games excite your dog to the extent he is moving so fast that he may jeopardise his recovery, slow him down by allowing him to search while wearing a harness and lead so that you can better guide and control his movement.*

CHEWING
Many dogs enjoy chewing well into adulthood and old age; it's not just an activity for puppies although, of course, their desire to chew is certainly strong. Appropriate chew toys and treats can help occupy your dog, give his teeth and gums a good clean, and even help reduce stress by relieving frustration.

A range of chews are available to suit all tastes (see image, right) and, for serious chewers, there are moulded plastic chews – for example, NylaboneTM – which are designed to withstand intense chewing.

If you are struggling to interest your dog in chews, try smearing the surface with something tasty, such as paté or soft cheese, or, if it's a rawhide chew, soaking a corner in water for a few minutes until it softens may encourage interest.

A range of suitable chew toys and treats are available, although you may need to try several before you find the ones that will keep him occupied for long periods of time. Many food toys also act as chew items, but it's good to have alternatives available.

Natural chew items include deer antlers and dried animal organs such as ears, lungs, heart, and tripe. Some rawhide chews – especially those with knots – can be chewed into smaller pieces and then swallowed whole, which can be dangerous, so choose your chews carefully, supervise initial introduction of new items, and remove any smaller pieces as necessary. Avoid anything that is brittle and can splinter into sharp shards as these can cause injury. Also available are ranges of manufactured edible chews, although these are often not as long-lasting as some of the natural chews. Many contain specific ingredients to help promote healthy teeth and gums.

No walks? No worries!

As with his toys, to keep your dog's interest in his chews do not leave the same ones lying around all the time. Have a toy/chew box, and rotate several at a time so that there is always something 'new' in the box. Dogs like novelty, and will show renewed interest in something that has been unavailable for a few days.

Remember!

Never use soft plastic toys, sticks or logs as chew items as these can easily splinter and be swallowed, leading to internal blockages or damage. Check all your dog's chews frequently, and remove or replace any that are worn or damaged, or any which have been chewed so small that they may be swallowed.

Tricks and treats

Learning new tricks is a great way to occupy your dog, and for the two of you to have fun together. You can also use these sessions to teach your dog useful skills, such as to tolerate or enjoy being handled for things like nail clipping.

Teaching him to go to a mat or a bed when he is asked can be very useful, whilst teaching him to 'play dead' by lying on his side when you say 'bang' can be a fun party trick, as well as encourage calm or still behaviour. For dogs who are more active, learning to turn in a circle clockwise and anti-clockwise, or weave through your legs, can be great fun. The possibilities are limited only by your dog's range of movement and desire to play this new game, though it is important always to ensure that you are not asking him to perform an action which may jeopardise his recovery or long term health, or repeat an action too many times in succession for the same reason. In in doubt, check with your vet first.

Several good training books exist which give step-by-step instructions on how to teach various tricks, and the Resources appendix has a list of suggestions. However, the key things to remember are –

- At first, train in short sessions with few distractions.

- Begin with easy actions, and build complexity over time.

- You may wish to use food to encourage your dog to perform the action you want at first.

- Reward your dog immediately he performs the action you're teaching.

- Break the training into small chunks if necessary.

What you teach your dog will depend on the nature of his physical restriction but can include touching his nose or paw to your hand, or another target such as a piece of sponge – as in the image here. Once he learns to touch the hand-held target, move the item to various places, so that he learns to touch a door (which he can learn to close) or (left) a flower pot (which may be useful to encourage him to hold his paw in one place for you to inspect).

- The cue word or hand signal (eg 'nose') should be added right at the end of your training, once the behaviour is perfect.

TIP
➤ *A cue, or command, is a verbal or visual signal which prompts your dog to perform an action or behaviour, eg: 'sit' means 'place your bottom on the floor.' The word or signal you choose has no meaning to your dog until he has associated it with a particular behaviour. Often, you will use a visual signal at first to prompt the required behaviour, eg: a wooden spoon moved in an arc to encourage your dog to turn in a circle. Teach the behaviour you want without a verbal cue at first, and only introduce the verbal cue when the behaviour is perfect.*

To add a verbal cue, say the cue word immediately before giving the visual signal. Your dog will begin to anticipate that hearing the verbal cue predicts seeing the visual cue, and therefore performing the behaviour. You can fade the visual cue once you start to see him perform the behaviour as you say the cue word.

A note on overstimulation

As well as being mentally occupied, dogs also need time to rest and sleep during the day: up to 14-18 hours a day in some cases, and possibly even more if recovering from major surgery or illness. Try to plan your dog's activities so that he has some fun and mentally challenging activities early in the day, and then a period where he can rest more quietly, with something to chew if necessary, or sleep if he wants to. This pattern can then be repeated later in the day. Too much activity can be as stressful as too little, and can sometimes lead to your dog finding it difficult to switch off and relax. If you think that this may be a problem and that he should be spending more time relaxing, look at his daily routine and try to re-organise things so that he is encouraged to rest at certain scheduled times.

No walks? No Worries!

Many tricks start with teaching your dog to target an object, which can be anything: your hand, a wooden spoon, or even a post-it note. Dogs generally like to investigate novelty so ensure you have a few treats ready in your pocket or a training pouch before you present the spoon (or object you want your dog to target) at nose height. As your dog moves to check out the item you've presented, mark that behaviour with a 'Yes' and give a food reward. Remove the object and prepare to present it again when your dog has finished eating, always marking and treating the correct behaviour.

Repeat up to ten times in one short training session, and you should very quickly have a dog who targets the item every time you present it.
When your dog is moving a few paces towards you to target the spoon when you present it, start to use it to encourage him to turn clockwise or anti-clockwise. Present the spoon and slowly move it in an arc. As your dog moves with it, say 'Yes' and reward him. You may not get as far as a full turn at first, but you can gradually increase the number of paces your dog moves, then mark and reward as before.

Worksheet 4: All about mental stimulation

What toys and chews do we need to buy/make? *(eg: I need to buy a couple of food dispensing toys, and ask friends to save food boxes for me so that I can put treats in them for him to rip up and get to the food)*

....

What games does he already know that can be modified for limited movement?
(eg: he likes to look for his toys to bring them for a game. Instead of playing a physical game I will let him search for them slowly on his harness and lead, and reward him with treats when he finds them)

....

What training resources will help me? *(eg: I will look on the internet for ideas of suitable tricks that he can learn)*

....

Planning mental stimulation into his routine
(eg: all his meals will be fed from food toys, spread over four smaller meals throughout the day)

Week 1: ....

Week 2: ....

Week 3: ....

Week 4: ....

Week 5: ....

Week 6: ....

Week 7: ....

Week 8: ....

Comments on progress; things that need to be amended
(eg: week 1: he doesn't want to engage in training after ten minutes so I will reduce each session to five minutes, and give him a toy or chew instead to occupy more of his time)

....

....

5 Emotional support for your dog

Alongside physical recovery, we are increasingly aware of the importance of supporting the mental and emotional needs of our dogs during a rehabilitation period. Dogs are a social species, and going for walks usually involves more than physical activity: it's a time to be sociable with people, other dogs, and even other animals, and therefore restricting walks can remove important social contact, too. As you plan for his period of restricted activity, this chapter will help you to keep this in mind, as well as other aspects of your dog's emotional wellbeing, namely –

- managing attention-seeking behaviour

- managing a dislike of veterinary visits

When identifying your dog's needs for emotional support consider how many people, dogs and other animals he meets on a daily basis, and whether or not these interactions are positive for him. Does he enjoy them, or would he rather not greet every dog he meets, or be hugged by the child from over the road? Like us, dogs are not automatically friendly with everyone, so think carefully about how he responds to interactions: with excitement and enjoyment or with tolerance, or even an attempt to avoid them? For dogs who are not that sociable, a period where their social interactions are more restricted may actually be relaxing for them, although it is important to ensure they do not become completely withdrawn, and find returning to more usual levels of activity emotionally stressful.

If your dog is fearful of other dogs, people or other animals, you may find he is more overt in his expression of this whilst he is recovering. If he is feeling more vulnerable, or has residual pain, he may be less tolerant of others approaching. Try to take this into account when planning outside trips or visitors, and ensure that everyone gives him more space, and respects his need for increased distance so that he is not forced to defend his personal space.

For fearful dogs, or dogs who take time to feel comfortable with new people or dogs, it is important that they maintain contact with dogs and people they do like and can relax with whilst their usual walks and social encounters are restricted. This is especially important if they will be limited in their activities for several weeks, or if they are puppies or adolescents.

No walks? No Worries!

Tip

❧ *Take extra care with respect to social interaction for dogs of up to three years old, even if they are not frightened or anxious about social encounters.*

Try to arrange for people to visit who will enrich your dog's day; who he will be pleased to see and can relax with. Being able to spend time with people outside your immediate household will provide a novel experience during his period of crate rest; even more so if the visitors are able to take part in some of his training or mental stimulation by playing appropriate games with him. Of course, as mentioned previously, this must be done in a manner that is positive and not stressful. If he travels well then you can also arrange to visit friends if it's possible (and their house is suitable for him in his current state of health).

continued page 64

Remember!

As important as it is to maintain positive relationships between your dog and others when he is ill or injured, a dog who is experiencing pain in the presence of another dog or a person – for example, if a boisterous dog inadvertently bumps into his sore leg, or someone strokes him over-enthusiastically – can form an association in his mind between the dog (or the person) and pain. Such an association can lead to him fearing interactions with other dogs (or people) in the future, so by carefully managing his interactions you are protecting not only his physical but emotional wellbeing, too.

Highly sociable dogs who are distressed when they are not allowed to interact will need different support during their activity restriction. If your dog is one who greets everyone and every dog as a long-lost friend, or who barks and lunges on-lead when denied an opportunity to interact (as in the photograph above), then the prospect of keeping him calm and quiet without regular social interactions may seem daunting. If you have time to prepare in advance for restricted walks, focus on teaching him a calm greeting so that when you have visitors he can greet them without risking injury to himself.

When he is once again allowed to stand and walk (in a controlled manner), consider whether he has any doggy friends who will greet him calmly and allow him to maintain some contact without getting overly excited or playful. At all times remember to protect him from any negative experiences. If you observe that he looks uncomfortable or is attempting to avoid an interaction, respect his communication and step in to remove the other dog or the person from his space.

It is also important to teach him how to cope when he is not allowed to greet someone or another dog. To teach this, or to manage a situation when you are outside and you think he is likely to get over-excited, keep a good distance away from the trigger (person, dog, etc) to reduce his excitement. Get his attention focused on you with a particularly tasty treat or a favourite toy, and use the food or toy to keep his attention and entice him further away. If you are able to practice this with friends, you will see him begin to associate the arrival of a person with something good coming from you, and may automatically begin to look to you in anticipation of the reward. If he does, reward him immediately and keep moving him away. If you decide he can greet the person or dog, use the food (keep feeding into his mouth, or drop treats on the ground for him to find) or a toy to keep him calm and focused on you, or the ground, as you approach the trigger. He is then more likely to greet them calmly as he hasn't been straining to get to them.

Try to ensure you give your dog sufficient opportunities to be with people and other dogs. Ensure that meetings between dogs remain calm and do not lead to frustration. You can see in this picture that an interaction is being carefully managed through the use of treats to keep the older dog calm whilst the pup looks on. Later, if both remain calm, allowing them to gently 'say hello' may be possible.

If you have visitors to your home and your dog becomes frustrated at being in his crate, ensure he has something he can rip or chew on to help reduce his frustration. Cardboard boxes filled with newspaper and a few treats are a useful tool at these times. If your dog likes to squeak toys then ensure he has one available (there are ultrasonic dog toys, apparently audible only to dogs, available from Hear Doggy™ if you find the squeak unbearable after a while).

It is natural to feel sorry for your dog when he is injured or unwell, and to give him more attention than he normally receives, especially if you have arranged your timetable to ensure he is not left alone at home for long periods while he recovers. He may also want more attention, or learn quickly that whining, barking or howling gets your attention, particularly if he wants you to continue playing a game or giving him a fuss. The best way to avoid or reduce any attention-seeking behaviour is to proactively ensure that he is able to settle and relax by himself (see Appendix 2 *Relaxation and crate training*) so that he is able to switch off. In addition, provide him with as much mental and physical stimulation as is appropriate (see Chapters 3 and 4) as part of his daily routine, scheduled to make best use of the times he is naturally more wakeful, as discussed earlier.

If you find your dog is beginning to vocalise, or scratch or chew the crate to get your attention, look at the routine you have established and see if you can make any changes which will pre-empt the attention-seeking. For example, dividing his daily meals into additional food toys spread throughout the day may help to keep him busy over a longer period. In addition, if you have left him alone to rest and relax and he begins trying to get your attention before you return to engage with him again, the most important thing is to ensure he's not successful in getting attention when he's doing something you don't like. If you are in the room with him, remember that any kind of eye contact or acknowledgement – even reprimanding him – is rewarding the attention-seeking behaviour. A good tactic is to immediately get up and leave the room without saying anything or even looking in his direction. Most dogs will quickly learn that vocalising has

the opposite effect to the one they want (as long as everyone is consistent in this approach).

If the noises and demanding behaviours don't stop quickly (or if you are out of the room when they start, and he doesn't stop of his own accord and give you the opportunity to approach him when he is quiet again), then make a noise – such as opening the fridge or back door, or the dog food cupboard – which will distract him, and usually result in him quietening down. Wait a few seconds while he is quiet and then return to the room. As long as he remains quiet, approach the crate and ask him to sit or perform another simple known behaviour, then reward him with attention. If he starts to bark again as you re-enter the room, leave immediately. You may need to repeat this a few times in quick succession in order for him to fully understand that barking makes you leave and being quiet gets him your attention.

Many dogs suffer separation-related problems when left alone. If your dog is predisposed to develop, or already has, these problems, be particularly careful if you are spending an increased amount of time with him while he is recovering. As soon as it is

To help your dog differentiate between when you will and won't interact with him, teach him a 'finish' cue which tells him that the interaction with you is over. When you have been stroking him or playing with him for a short period, withdraw your attention (you may find a large visual cue such as holding up both hands, palms facing him, works best) and say 'all finished,' then give him a chew or scatter a few treats on the ground and walk away from him. The visual and verbal cue will soon be associated with the arrival of the treats – something to keep him occupied on his own – and the withdrawal of your attention ...

When he has finished the treats or chew, but before he has begun to ask for your attention again, go back to him and offer him the chance to interact with you. This also helps teach him that being calm and relaxed gets your attention. Finish the encounter as before. Ensure all the family follow the same rules to help him learn this. Over time you will be able to fade out the treats and simply tell him 'all finished,' and he will have learnt to cope with withdrawal of your attention.

safe to do so, ensure you continue to leave him alone for periods of time, keeping your comings and goings as neutral as possible. If you are particularly concerned, try to leave him with something good to keep him occupied, like an appropriate stuffed food toy, and make your initial time away brief.

TIP
🐾 *You can video record him whilst you are out to see exactly how he's behaving, and then adjust your behaviour accordingly. If you have concerns about his behaviour, discuss this with your vet as soon as possible, as referral to a suitably-qualified behaviour practitioner may be necessary.*

Recovering from an illness or injury can leave your dog feeling vulnerable, and/or with a memory of being in pain, especially when handled or when at the veterinary clinic. This learned expectation of pain can cause him to attempt to avoid being handled, by using aggressive behaviour or escape tactics. He may even become distressed when approaching the veterinary surgery or in the consulting room, because he has learned that

being there is unpleasant, painful or scary as a result of the unavoidable handling and treatment he has received. Discuss with your vet the possibility of scheduling visits where he has no treatment, but can slowly become comfortable again in the clinic. Most vets are happy to help their patients relax in their surgeries.

If your dog is really struggling with the clinic environment, in extreme cases your vet may be able to examine or treat your dog in the car park, or provide a sedative that will relax him before your visit. Some vets may be able to offer a home visit, although this will depend on why your dog needs to go to the vet, and whether or not the vet needs any specialist equipment or support available in order to appropriately examine or treat him. It's also important that your dog does not begin to make an unpleasant association between his home (where he should feel most relaxed and secure) and uncomfortable or scary veterinary examinations or treatment. If dogs are treated at home, the treatment must never take place in their bed or other safe haven. It is important to discuss any concerns you have regarding vet visits with your vet as soon as possible so that problems can be addressed early.

(And overleaf): Help your dog to love your vet by making regular visits to the surgery simply to sit in the car park or waiting room and give him treats. When you see him begin to show positive anticipation when heading into the surgery – tail and head up (rather than tucked), relaxed or happy expression on his face, eager to get into the surgery – then you can progress to taking him into the consulting room, giving him time to relax, and feeding him treats.

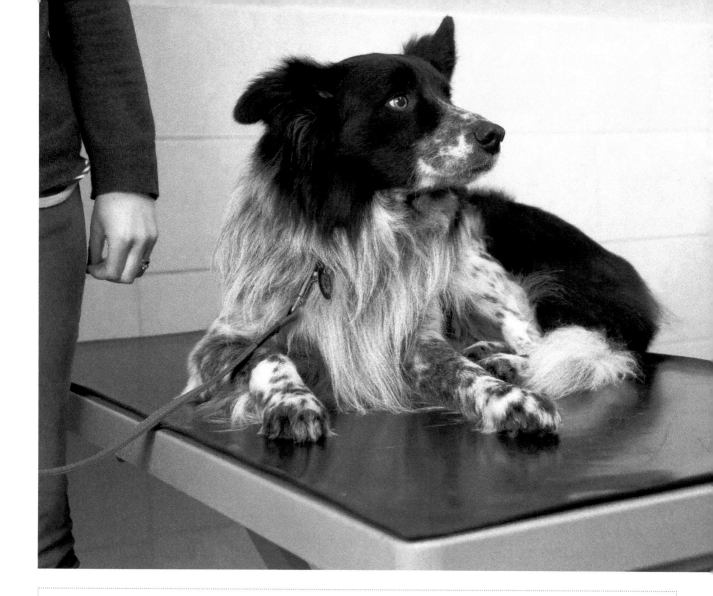

REMEMBER!

If in pain, fearful, or frustrated, your dog may snap at or bite someone during treatment or recuperation. If this happens, the first thing to remember is not to react negatively and scold or punish him. Try to remain calm and remove him, or the person he bit, from the situation as gently and quietly as possible with minimum fuss. It is a very shocking thing to happen, and it is natural to be upset, or angry; however, punishing a dog who has bitten will often only serve to increase his anxiety about the situation, and potentially increase the chance of it happening again. It's a good idea to discuss the situation with your vet as soon as possible as referral to a suitably-qualified behaviour counsellor for specialist help may be necessary.

WORKSHEET 5: EMOTIONAL SUPPORT FOR YOUR DOG

Current considerations regarding his general outlook on life *(eg: he is a very sociable dog who will probably miss interaction with others)*

....

....

Plans to manage social interactions *(eg: I will invite friends round at least once a week to spend structured time with us so that he can maintain contact with people outside the household)*

....

....

Plans to manage attention-seeking behaviour *(eg: I will have a stash of chews close to the crate at all times so that before he has learnt to tolerate being confined away from me I can easily give him a chew if I have to leave him for a period of time)*

....

....

Plans to manage vet visits and handling *(eg: I will spend some time every day gently touching different areas of his body, pairing this with food treats to help him recognise and remember that being handled is good)*

....

....

Notes on progress and special considerations *(eg: he is totally relaxed with me even when I touch him around his sore area. I will now ask friends to participate in handling sessions (as long as he remains relaxed and comfortable) so that he learns to trust other people touching him, too)*

....

....

....

69

The day your vet says you can begin taking your dog for walks again will probably come as a relief, no matter how well you and he have adjusted to your temporary routine. However, try not to let your excitement take over when planning his new routine, as it's best to reintroduce walks and additional physical activity in a controlled way, whilst continuing to include many of the alternative games and training you have been engaging in whilst he's been recuperating.

Depending on how long he's been on restricted walks, and how much replacement physical activity – such as hydrotherapy – he's been able to do, your dog may have some muscle wastage as a result of his confinement, as well as potentially a loss of fitness. If you have not already consulted a canine physiotherapist, now may be a good time to do so to draw up a programme of massage and stretches which can help alleviate muscle stiffness and soreness as exercise increases.

To reduce the chances of further injury during this transitional period, consider a phased reintroduction of walks of gradually increasing duration and difficulty. Also, on each walk, include time for him to warm up with some gentle on-lead walking before letting him off-lead to manage his own activity levels.

TIP

❥ Every dog will have a different response to that first off-lead opportunity, post-exercise restriction. Some will continue to potter slowly, whilst others will take full advantage of that first moment of freedom, and set off full pelt across the field. If you suspect your dog may do the latter, let him off in an enclosed space at first, with a good supply of very tasty treats that you can use to gain his attention before he risks injuring himself, or undoing all the good work of his recovery.

Always finish each walk with a few minutes of slow, on-lead walking to help reduce any stiffness.

For the first few weeks alternate short walks – which include off-lead time – with days of on-lead only walks, and be guided by your dog's responses and behaviour (during and after

If you have a dog who used to play games (such as fetch or tug) on walks, do not reintroduce these activities immediately, but rather build up a level of fitness and muscle strength before introducing the bursts of speed, fast stops and muscle tension that these sort of games involve. Of course, some injuries or surgery may mean stopping these games entirely, or changing the way you play them to minimise risk.

If you see that your dog's arousal levels are increasing (hackles are rising, body is stiffening, tail and ears becoming more erect, for example), use a food treat to divert his attention and move him slowly and calmly away (as described in the previous chapter).

the walk) when considering whether to increase, decrease, or maintain the level of activity. It's possible that increased exercise may result in additional pain for him, so look for signs of stiffness or reluctance to move (for example, when jumping into the car or onto the sofa, if this is once again allowed). You may also see behavioural indications of increased pain such as avoiding being touched. If this happens, discuss increasing pain relief with your vet, and reduce exercise again until the pain is under control.

You may notice that your dog shows a heightened awareness or alertness the first few times he is again allowed out on walks. If, during his recuperation, you have been able to take him for drives, or places outside his garden where he can relax, hopefully, this hyper-vigilance will be mitigated somewhat, but it's something to be aware of. It can be motivated by increased fear if he was already a fearful dog, or if he feels vulnerable because of his illness or injury, or the cause may be heightened arousal

To help him cope with reintroduction to the wider environment, give your dog plenty of distance from things that appear to cause him to become alert, as well as time to assess situations and take things in. In many cases, simply being able to evaluate a situation will allow him to turn away and remain calm.

and frustration at being restrained when he wants to be free. In some cases, it can become a learned response because it is inadvertently rewarded: for example, by getting a lot of attention for the behaviour if he is constantly reassured when scanning the environment.

If you have used a food treat to move your dog out of a situation where he is becoming aroused or agitated (as described in the caption on page 72), at a distance where he becomes calm, allow him to look again at the thing bothering or exciting him, and look away in his own time. If he remains calm, reward him as soon as he does anything other than stare at whatever's concerning him. If he wants to, move a few steps closer and allow him to look again; if he chooses to walk away then go with him. Do not tempt him toward something that is bothering him, or encourage him to get closer than feels comfortable with, but reward him if he approaches of his own accord (if it is safe to do so), and remains calm. You should see a decrease in vigilance and a reduced tendency to react to things as your walks become more frequent again. If this doesn't resolve on its own within a few weeks, consulting an appropriately-qualified behaviour specialist is advisable.

As you gradually reintroduce your dog's walks, take time to reassess his needs on an ongoing basis. You have probably been more than usually aware of his emotional state during this time of restriction, and have a good idea of the things he has really enjoyed or benefited from. The changes you have made, and the shift in emphasis away from daily walks to different forms of mental, physical and emotional support, are potentially beneficial

in their own right. So, for example, it may be advantageous, even in the longer term, to reduce his walks and replace a couple with scent games, or introduce some scent games to your daily walk to change their emphasis. If he has really enjoyed earning his meals learning new tricks, or being fed via food toys, continue to do so rather than reverting to meals from a bowl every day.

If you are facing a longer term adjustment to his physical activities – for example, if his medical condition requires permanent changes in activity, or if his walks are restricted for behavioural reasons – take time yourself to adjust to this state of affairs. Accepting the dog you have now, rather than the one you used to have or wanted to have, can take a while, and it is natural, initially, to feel anger, disappointment or denial at the situation in which you find yourselves. The range of activities described in the preceding chapters will have provided alternative ways for you to enjoy each other's company, but be realistic about what he will be able to do in the longer term.

Hopefully, now that you have reached the end of the process of adapting to a change in your dog's activity levels, using the approach described here has helped to reduce stress for you and for him. Throughout this process you have assessed him as an individual, and acted to ensure his needs are met in the best and most appropriate way for your circumstances. Continue to do this as his needs evolve throughout his life, and enjoy the relationship which accepting one another's abilities and restrictions brings.

If you and your dog have been involved in canine sport, and wish to continue with organised activities, consider changing to those which put less stress on the body – tracking, for example – which can be made more difficult mentally by tracking over a range of surfaces, such as tarmac, gravel, and even through buildings, without adding physical stress. Other scent training activities are also possible, and some trainers offer their own accreditation system to provide goals to work towards. It may also be worth contacting trainers in your area to see if they offer classes that are designed to teach fun games and skills that challenge you and your dog mentally, rather than putting a great deal of emphasis on physical activity. Of course, always assess any training to check that it suits your dog's requirements and capabilities; that the environment is comfortable for you both, and that the training methods used are focused around positive reinforcement.

Worksheet 6: Back to normality

PLANNING
What are the possible problem areas when walking resumes? *(eg: over-excitement on first walk; fearfulness when meeting other dogs, etc)*

....

What have we put in place to avoid this? *(eg: ensure first walk is mostly on-lead, and he is only allowed off-lead in a restricted space so he can't run much)*

....

How often/long will we walk, and where?

Week 1: ....

Week 2: ....

Week 3: ....

Week 4: ....

Week 5: ....

Week 6: ....

Week 7: ....

Week 8: ....

NOTES (AFTER WALKS HAVE RESUMED)
Problems not predicted *(eg: he reacted fearfully to a dog he used to play with)*

....

Plans implemented in light of these observations *(eg: keep a greater distance from other dogs in the short term, and gradually allow him to move closer as he becomes more comfortable to do so)*

....

....

APPENDIX 1 Loose lead walking

Walking on a loose lead comes naturally to some dogs; others can take a while to master the skill of walking at our pace, and dividing their attention between the world of sniffs and scents around them and the human at the other end of the lead. There are many different ways to teach dogs to walk nicely on a lead, and, as long as the technique you choose is not detrimental to your dog's welfare, it's up to you to determine the method that best suits you both.

It is likely you've tried various techniques before, but if you are struggling it is worth starting again from the beginning – if necessary, swap the side that your dog walks on to make it different for him – and being consistent in your approach.

There will be times when you just need to get from A to B, and don't have the time or the patience to be consistent with loose lead walking, so a different harness or collar is a good way of differentiating this for your dog. We suggest you pick the collar or harness you want to use when you walk with your dog on a loose lead, and then use a different type for the times when he is allowed to pull: for example, use a harness with an attachment at the back for loose lead walking, and a front attaching harness to walk him at other times. The following is a suggestion for a training technique which we find works well: an amalgamation of a number of different ideas learnt from numerous sources over the years.

The aim of this training technique is to enable your dog to walk with you without pulling, whilst still enjoying the walk and the environment around him. Or course, you should both enjoy the walk as we regard loose lead walking as a partnership between person and dog: a contract in which your dog's undertaking is not to pull and not to trip you up, and yours is to allow your dog to enjoy the walk by doing the things he likes – sniffing bushes, investigating holes, and greeting people and dogs where appropriate. This technique is not intended to teach obedience heelwork.

- Begin with an appropriately-fitting flat collar, harness or head halter attached to a lead, which is long enough to allow your dog to walk comfortably alongside you with slack in the lead between your hand and him.

- Decide on which side you would like your dog to walk.

- Arm yourself with tasty, easily swallowed treats that are cut up very small – eg: cheese, chicken, frankfurter.

- Find a quiet, distraction-free, safe environment in which to work. Your living room or garden are ideal places to start at first.

- Attach the lead to your dog's collar or harness and stand still.

- Say nothing, and, as soon as he pays you any attention, praise him and feed him a treat.

- Now use a treat in your fingers to bring him into position at your side.

- Make sure the lead is loose and that there is no pressure on his collar/harness: the lead is simply a safety line not a means of directing his movement.

- Feed him treats simply for standing next to you. He is learning that being at your side is a good place to be.

- Take one step forward and, as he moves with you, feed him a treat into his mouth. Try to feed him for staying alongside you; don't wait for him to get ahead. If he is reluctant to move as you step forward, use another treat to entice him into position next to you again.

- If he moves along keeping pace with you, feed him a couple more treats as you slowly take a few steps forward.

continued page 80

Before starting to walk, reward your dog for simply standing next to you and paying attention to you. Take a step forward and reward him frequently (every step or two) for walking next to you and paying attention to you. Keep your hands up by your waist so you are not enticing him with food.

Should he pull ahead of you, stand still, and don't begin moving, even if he pulls hard on the lead.

(Both pages): In the early stages of training use a treat to help your dog move back to your side. Later, use just a hand signal, and eventually expect him to be able to move back of his own accord.

- After practising this for a few days, walk a little further before feeding a treat, so that it isn't quite as easy for him.

- If, at any time, he gets ahead of you or becomes distracted and pulls sideways or backwards, stand still

immediately, don't say anything, and wait for his attention to return to you. Once you have this, use a treat in your fingers or a hand gesture to return him to your side, and have him walk a couple of steps next to you before receiving a treat.

He doesn't have to be glued to your leg, just moving with you and not pulling.

Take care not to feed him a treat immediately he comes back to you as this can very quickly teach him that the quickest way to get a treat is to pull away and then return.

- Repeat this initial training in different places, with gradually increasing distractions. Practise until he can walk alongside you for about 20 paces without pulling before receiving a treat, and if he does pull, can return to position to be able to move forward without requiring a hand signal to prompt him.

No walks? No Worries!

Once he is back at your side and paying you attention, begin moving forward again.

- At this point, for most dogs, moving forward becomes a more important reward than the food, so you can try and fade out the food rewards, unless in a very distracting place where he will need extra help to get it right.

- When you reach this stage of training, you may find that he develops the very annoying habit of yo-yoing on the lead: pulling and then coming back so that your walk becomes a stop/start affair. In this case, the next part of the training becomes very important.

- Next time your dog pulls and then moves back in to position, DON'T continue forward. You'll probably find that he anticipates you're going to move and bounces forward to pull again. Stand still.

- At this point you want your dog to understand he has to be with you mentally before he can move forward. In other words, he has to be paying you real attention, and be in tune with you, to be able to move forward, not focusing totally on the environment and just flicking you an occasional look.

- Judging this mental state can be difficult, but look for signs that he has moved his attention from the environment and firmly back to you. Some dogs demonstrate this by sitting next to you; some show a more relaxed posture instead of being intently focused on the environment; some make sustained eye contact with you.

- When you see this change in behaviour, tell him he is very good and move forward again.

- It may be beneficial to use food rewards again for a while so that the change in mental attitude receives two payoffs – food and movement – but DON'T give in to the temptation to use a food treat to move him in to position. At this point we want your dog to take responsibility for walking without pulling, which means that it will no longer be necessary for you to manage him or remind him as he will have learned to manage himself.

- Over time he will develop the skill of paying you just enough attention to match his pace to yours, and change direction with you while enjoying his surroundings. At this point he won't be actively looking at you at all (or rarely) but will simply be keeping you in his peripheral vision.

Over time he will learn to divide his attention between you and the environment; matching his pace to yours and resisting the temptation to pull. Make sure you give him opportunities to sniff and check out things on your walks, too, by taking him to potentially interesting places, or places he seems to be paying some attention to, if it's appropriate.

Appendix 2 Relaxation & crate training

Many dogs have to learn how to mentally relax as opposed to simply sleeping because of physical tiredness, and this is especially true when their physical activity is restricted. You can teach your dog to relax, both in and out of a crate, but the following description assumes you will use a crate, as this is frequently most useful with an activity-restricted dog. If you want to teach relaxation on a bed rather than a crate, the technique will work in the same way (see the pictures opposite for an illustration of this).

The same technique can also be used for teaching relaxation in the car, or any other space where you want your dog to feel comfortable, safe, secure and relaxed. The aim of this training is not to teach your dog to simply tolerate being in a crate (or whichever space you choose), but to create an association with calmness, relaxation and security within that space.

If you are revisiting crate training with a dog who has previous negative associations with a crate, try to change the way the crate looks and feels to your dog. For example, can you change the position of the crate; use an end door instead of a side door by which he enters it; cover it (or remove a cover if used before)? If you have previously used his bed as a time-out or punishment for bad behaviour, we suggest you buy a new bed specifically for his crate, and do not use the crate as a punishment area in future.

Undoing previous bad experiences may take a while, so in your planning allow plenty of time to prepare him. In an emergency situation, try to implement as many of the suggested

Relaxation training in a bed

1 Encourage your dog to investigate or go in to the crate or bed by placing a tasty treat, or his food bowl, in it.

2 Reward your dog for going in to the crate, or on to his bed, by dropping treats inside as soon as he moves towards it.

3 When your dog is happy to go to his crate or bed, begin to delay the treats to encourage him to lie down; reward him once he has done so.

4 Once he understands that laying down brings rewards, wait for additional signs of relaxation, such as laying with his hips to one side, dropping his head or closing his eyes before rewarding. If your dog is excited by the prospect of food rewards, you may find that gentle stroking or just calm verbal praise is a better way of reinforcing his relaxed behaviour at this stage.

changes as possible (to make the crate appear different), and expect to spend longer sitting by his crate as described later.

Choose a warm, cosy location for his crate: somewhere he can be part of the family but without too much passing traffic or noise. Avoid placing the crate next to a washing machine or in a busy passageway. Some dogs find an Adaptil™ diffuser helpful, and will relax more quickly if one is placed near to their crate.

Steps to crate training

- Choose a crate that is large enough for your dog to stretch out in, and to stand up and turn around easily. The dog shown on page 86 chooses to sleep in this crate, but if she were confined for a period of time she would need a larger crate.

1

2

3

4

No walks? No Worries!

- Set up the crate in your chosen location, place a comfy bed and something that smells of you (an old T-shirt, for example) inside, and leave the door open. Many dogs will take the opportunity to investigate the new object, so be ready with treats to reward any interest in the crate, dropping treats just inside at first, and gradually moving them further in as his confidence grows. If you have plenty of time for crate training, and no other dogs in the house, you can leave high value chews in the crate for him to find. After a few repetitions you'll see him begin to investigate the crate with positive anticipation of finding something good.

- Try to encourage him to walk in himself. If he does, make a huge fuss of him, feed him treats while you close the door, and continue once the door is shut.

- If you are short of time, tempt him into the crate with high value treats, or, if it is possible (particularly if he is not that food-motivated), make his first introduction at a meal time, placing his food bowl (with high value food added to his usual meal) inside.

- In an emergency, if he is reluctant to go into the crate, try to very gently lift him in (ONLY if this does not pose a risk to him or you), and then feed him treats through the closed door.

- In an emergency, if it is evident that he won't go in without the risk of aggression, consider using a less restrictive means of confinement in the beginning. For example, you can put matting down in a bathroom or utility room and place his bed on top of that. This is usually a small enough space to prevent excessive movement. If he struggles to cope with the door closed, fit a stair gate in the doorway so that he can see out. Of course, these spaces can be used long term, although they are not usually somewhere that he can still be part of family life, and could be inconvenient in the long term.

- The first time you close the door of the crate, plan to sit next to it for a while to keep him company, and feed him treats or give him a stuffed food toy or chew to keep him occupied and settled. If training in preparation for future confinement, open the door before he has finished his chew, and let him choose when to leave the crate. In an emergency, once he is engaged with the chew or toy, you can move away from the crate, but always return before he shows signs of becoming agitated or distressed. You are trying to avoid him becoming frustrated or anxious about being in the crate, so at this stage you need to stay with him and provide reassurance by your presence, or entertainment through food toys and chews. You will be able to leave him for longer as he adjusts and learns how to relax by himself.

- When training in preparation for restriction, leave the door open at all times, and reward him any time you see him heading in to the crate. Close the door when you give him a food toy or chew in there, or you have a few minutes to practice relaxation in the crate. At first, choose your practice times so that he is likely to be relaxed or tired: for example, after a walk or a training session.

- He will probably be quite alert to start off with, sitting or standing in the crate, watching for the treats to arrive. Deliver the treats in a slow, gentle way, directly on to the bed in front of him. Try not to ask him to lie down, but encourage it by the way you place the treats. You can wait a while between treats and see if that delay prompts him to lie down. As soon as he does lie down, deliver a treat directly to him again. Gradually increase the delay between treats, and begin to reward signs of relaxation, such as his head dropping, moving on to his side, or even deep breaths or sighs.

- As he gets more practice – and reward – for being relaxed in his crate, begin to wait for him to relax before offering the first treat or chew. Remember, it's not just a case of him lying down; you are looking for signs of emotional relaxation, too. If treats are too arousing for him, use gentle strokes – if you can reach him through the bars of the crate – or just a relaxing tone of voice and interaction with you to reinforce the behaviour. If necessary, remove his bed from the crate and teach this on the bed first, so you can stroke him if that is reinforcing for him, and then return the bed to the crate to continue with the training.

Right: This dog is truly relaxed in the crate. As you prepare for the period of restricted activity, gradually increase the amount of time he spends in the crate, starting off by encouraging him to rest and sleep in there at his usual rest times.

No walks? No worries!

You will know your dog is comfortable, safe and happy in his crate when he chooses to relax in there when the door is open, even though there are other options, such as the sofa, available to him.

- Slowly increase the amount of time he spends in the crate as you transition towards the routine you will follow once his activity is fully restricted.

- Only open the door to his crate if he is calm and relaxed. It is normal for him to become excited at the prospect of being allowed out – even if it is only to go to the toilet – but wait for him to be calm before undoing the latch. If he risks injuring himself, use food scattered at the back of the crate to divert his attention whilst you open the door. Other tips to help with attention-seeking-type behaviour can be found in the chapter *Emotional support for your dog*.

- In an emergency, while he's adjusting to being confined, and to save distress and noise disturbance all round, sleeping alongside the crate may prevent him barking or howling out of frustration or because he's isolated. Trying to ignore vocalisation and leaving him to bark is a risky strategy because he may escalate his behaviour, and try to chew or dig his way out, hurting himself in the process, or his barking may continue longer than your ability to ignore it. If you then go to him, you will have taught him that barking works. Better to prevent the barking in the first place and slowly wean him away from your presence over time.

- We do not recommend using any form of punitive anti-bark device as part of crate training – your dog will be living in his crate for several weeks, and its primary function is as a safe, secure, relaxing and comfortable place to be. Using any form of punishment in this environment is counter-productive and potentially damaging to your dog's emotional wellbeing.

APPENDIX 3 Further information

Books

The following books may provide more ideas for continuing with general training as well as scentwork training –

The Complete Idiot's Guide to Positive Dog Training, Third Edition. Pamela Dennison

10-Minute Dog Training Games. Kyra Sundance

Brain Games for Dogs. Claire Arrowsmith

Smellorama! Nose Games for Dogs. Viviane Theby

The Canine Kingdom of Scent: Fun activities using your dog's natural instincts. Anne Lill Kvam

Websites

www.nowalksnoworries.com

Notes

Rogues' Gallery: The Cast

Frankie

Tyke

Spook

Dobie

uMoya

Izzy

Bramley

Toffee

HELPING MINDS MEET
Skills for a better life with your dog

Helen Zulch & Daniel Mills

Hubble & Hattie

This unique book, written by professionals in the field, explains how and why misunderstandings occur between us and our canine companions, and how we can work to resolve them. It aims to help us adjust the way we interact with our dogs, in order to help our dogs be well behaved, whilst at the same time enabling us to enjoy fulfilling relationships and a good quality of life with our canine companions

205x205mm • 96 pages • 100 colour images • paperback plus flaps • ISBN 9781845845766 • £12.99*

LIFE SKILLS FOR PUPPIES
Laying the foundation for a loving, lasting relationship

Helen Zulch & Daniel Mills

Hubble & Hattie

Puppy education from the puppy's perspective! Presenting the key skills that a dog needs to cope with life, this ground-breaking book, written by professionals in the field, aims to assist owners develop a fulfilling relationship with their puppy, helping him to behave in an appropriate manner and develop resilience, whilst maintaining good welfare. The skills taught are incorporated into everyday life so that training time is reduced, and practising good manners and appropriate behaviour become a way of life.

205x205mm · 96 pages · 121 colour images · paperback plus flaps · ISBN 9781845844462 · £12.99*

Index